Animal Groups

# Reptiles

By Dalton Rains

www.littlebluehousebooks.com



Little Blue House is distributed by North Star Editions:
sales@northstareditions.com | 888-417-0195

Produced for Little Blue House by Red Line Editorial.

Photographs ©: Shutterstock Images, cover, 4, 7, 9, 11, 13, 14–15, 17, 18–19, 21, 22, 23, 24 (top left), 24 (top right), 24 (bottom left), 24 (bottom right)

**Library of Congress Control Number: 2023902025**

**ISBN**
978-1-64619-812-2 (hardcover)
978-1-64619-841-2 (paperback)
978-1-64619-898-6 (ebook pdf)
978-1-64619-870-2 (hosted ebook)

Printed in the United States of America
Mankato, MN
082023

## About the Author

Dalton Rains writes and edits nonfiction children's books. He lives in Minnesota.

# Table of Contents

lizard
scale

# Reptiles

The lizard is a reptile.

It has scales.

The snake is a reptile.

It has no arms or legs.

snake

The sea turtle is a reptile.

It has a shell.

shell
sea turtle

The tortoise is a reptile.

It hatches from an egg.

tortoise
egg

The alligator is a reptile.

It is large.

alligator

The crocodile is a reptile.

It has many teeth.

crocodile
tooth

The Komodo dragon
is a reptile.
It has a long tongue.

Komodo dragon
tongue

The gecko is a reptile.

It has spots.

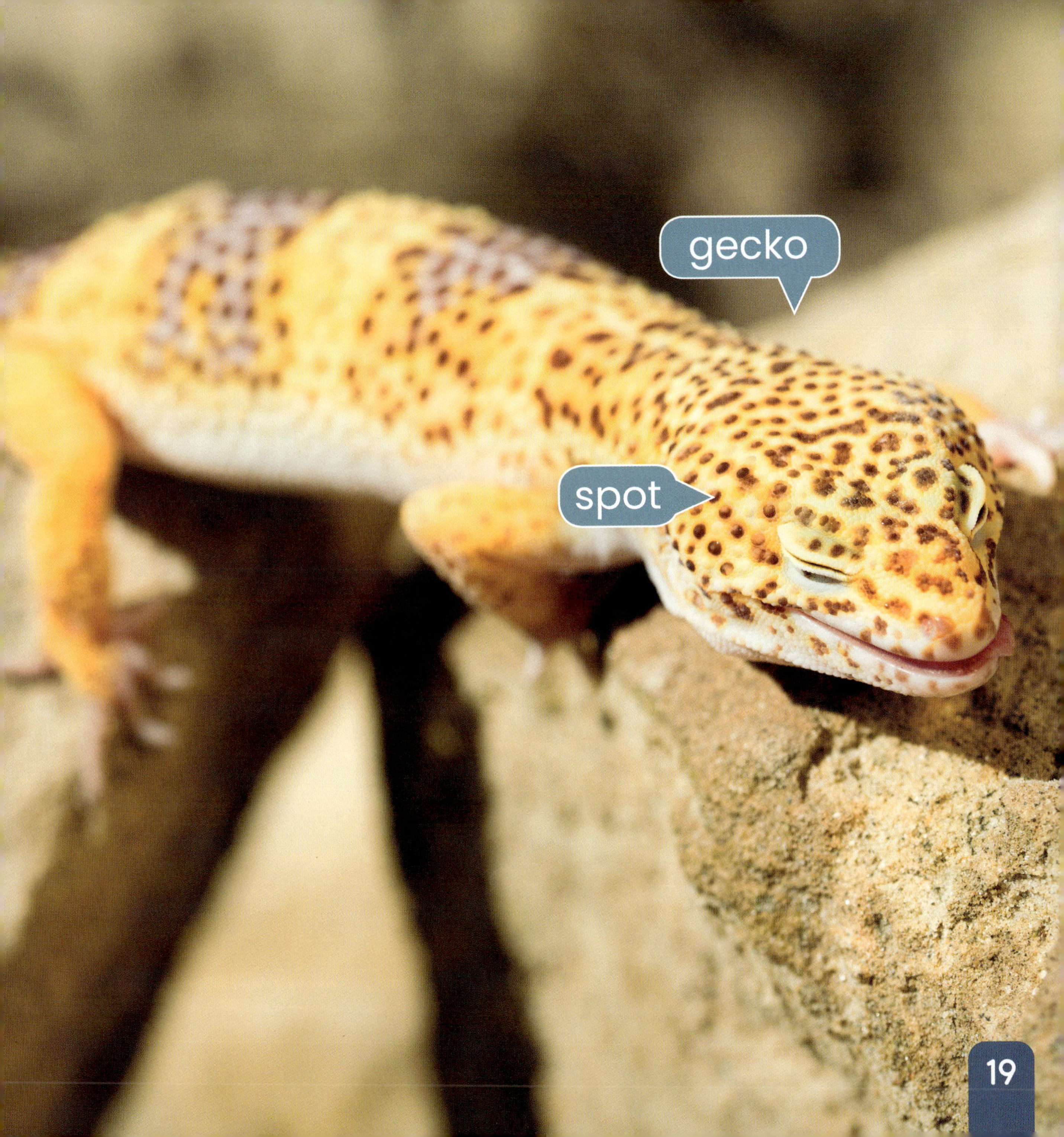
gecko
spot

The chameleon is
a reptile.
It eats bugs.

chameleon

# Is It a Reptile?

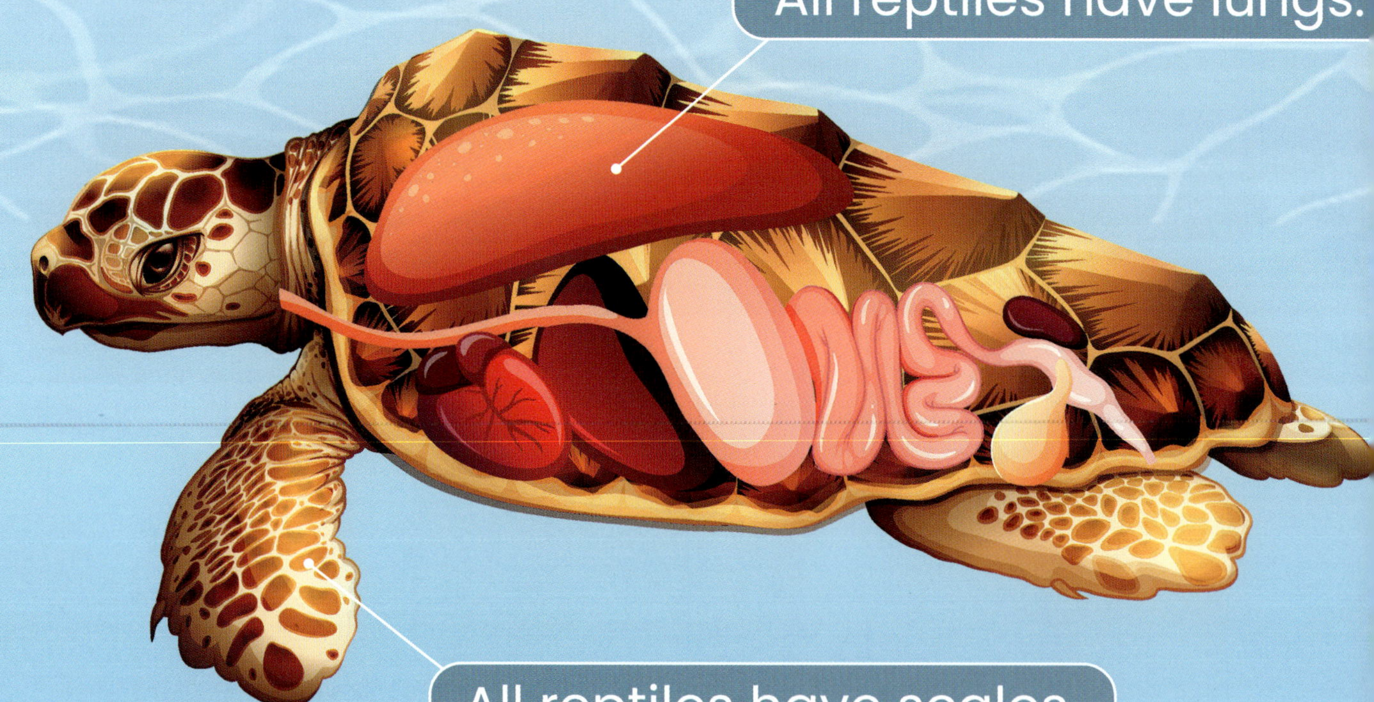

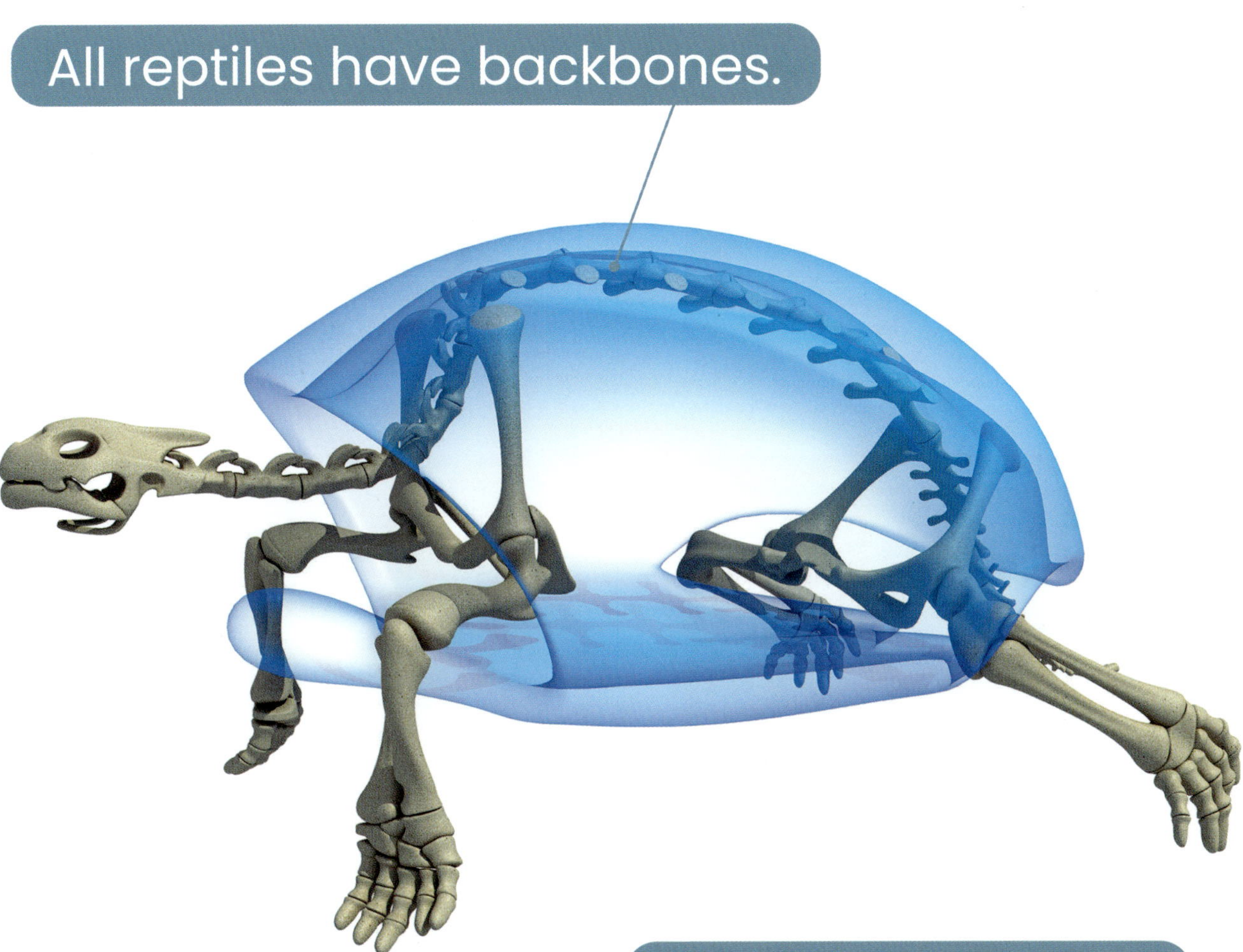


All reptiles have temperatures that match the outside.

# Glossary

**alligator**

**Komodo dragon**

**chameleon**

**sea turtle**

# Index